MIGHTY CONSTRUCTION MACHINES

EXCAVATORS

RYAN JAMES

A Crabtree Roots Book

Crabtree Publishing
crabtreebooks.com

School-to-Home Support for Caregivers and Teachers

This book helps children grow by letting them practice reading. Here are a few guiding questions to help the reader with building his or her comprehension skills. Possible answers appear here in red.

Before Reading:

- What do I think this book is about?
    - *I think this book is about how excavators work.*
    - *I think this book is about how excavators are made.*
- What do I want to learn about this topic?
    - *I want to learn what excavators do.*
    - *I want to learn the parts of an excavator.*

During Reading:

- I wonder why...
    - *I wonder why some jobs need an excavator.*
    - *I wonder why it is called an excavator.*
- What have I learned so far?
    - *I have learned that most excavators have tracks.*
    - *I have learned that some excavators are green.*

After Reading:

- What details did I learn about this topic?
    - *I have learned that all excavators dig into the ground.*
    - *I have learned that most excavators have a bucket.*
- Read the book again and look for the vocabulary words.
    - *I see the word* ***tracks*** *on page 7 and the word* ***ground*** *on page 8. The other vocabulary words are found on page 14.*

This is an **excavator**.

Some excavators are green.

Most excavators have two **tracks**.

All excavators dig into the **ground**.

Most excavators have a **bucket**.

How many excavators do you see?

Word List

Sight Words

a	green	most
all	have	see
an	how	some
are	into	this
dig	is	two
do	many	you

Words to Know

bucket

excavator

ground

tracks

30 Words

This is an **excavator**.

Some excavators are green.

Most excavators have two **tracks**.

All excavators dig into the **ground**.

Most excavators have a **bucket**.

How many excavators do you see?

MIGHTY CONSTRUCTION MACHINES

EXCAVATORS

RYAN JAMES

Written by: Ryan James
Designed by: Rhea Wallace
Series Development: James Earley
Proofreader: Melissa Boyce
Educational Consultant: Marie Lemke M.Ed.

Photographs:
Shutterstock: Eduard Valentinov: cover; magee: p. 1, 3; dani danier: p. 5; Another77: p. 6-7; Andy Dean Photography: p. 8-9; gyn9037: p. 11; 12-13 ABCSstock: p. 12-13

Crabtree Publishing

crabtreebooks.com 800-387-7650

Printed in Canada/012024/CP20231127

Published in Canada
Crabtree Publishing
616 Welland Ave.
St. Catharines, Ontario
L2M 5V6

Published in the United States
Crabtree Publishing
347 Fifth Ave
Suite 1402-145
New York, NY 10016

Library and Archives Canada Cataloguing in Publication
Available at Library and Archives Canada

Library of Congress Cataloging-in-Publication Data
Available at the Library of Congress

Hardcover: 978-1-0398-3840-6
Paperback: 978-1-0398-3925-0
Ebook (pdf): 978-1-0398-4008-9
Epub: 978-1-0398-4080-5